5th Engineering Cosmetics and Consumer Products Conference (ECCP 2024)

Philadelphia, Pennsylvania, USA
22 - 24 April 2024

ISBN: 979-8-3313-0080-7

Printed from e-media with permission by:

Curran Associates, Inc.
57 Morehouse Lane
Red Hook, NY 12571

AIChE
120 Wall Street, FL 23
New York, NY 10005-4020

Phone: (800) 242-4363
Fax: (203) 775-5177

www.aiche.org

Additional copies of this publication are available from:

Curran Associates, Inc.
57 Morehouse Lane
Red Hook, NY 12571 USA
Phone: 845-758-0400
Fax: 845-758-2633
Email: curran@proceedings.com
Web: www.proceedings.com

TABLE OF CONTENTS

KEYNOTE PRESENTATION

SESSION 1 - CROSS-INDUSTRY INNOVATION IN FORMULATED PRODUCTS

SESSION 2 - ENGINEERING SOLUTIONS FOR SUSTAINABLE PRODUCTION

SESSION 3 - REGULATIONS, SAFETY, AND COMPLIANCE

SESSION 4 - NOVEL MATERIALS IN HAIR CARE

SESSION 5 - SUSTAINABILITY IN HOME CARE

SESSION 6 - NOVEL MATERIALS IN SKIN CARE

Microfluidics to Macro Impact: Manufacturing Marvels

CONFERENCE

Conference on Engineering Cosmetics and Consumer Products

YEAR

2024

PROCEEDING

5th Engineering Cosmetics and Consumer Products Conference

GROUP

General Submissions

SESSION

Keynote Presentation

TIME

Wednesday, April 24, 2024 - 9:00am to 9:45am

Authors

Lee, D. - Presenter, University of Pennsylvania

Taking advantage of the precise control over the fluid flows in microchannels, droplet microfluidics has enjoyed a tremendous success in enabling synthesis of highly uniform and uniquely structured liquid droplets and gas bubbles. Although some successes in translating microfluidic-based assays and diagnostics to commercial technologies have been achieved, few examples of such translations in the domain of materials manufacturing based on microfluidics have been accomplished, largely due to the very low throughput of droplet/bubble generations. Without solving the scale-up challenge, microfluidic-based materials synthesis will continue to remain an academic exercise that may not reach its true potential in transforming various sectors of industry including medicine, personal/home care products, separations and catalysis. In this talk, I will present the recent developments that we have made in enabling AI-augmented control of microfluidics and large-scale manufacturing of materials using microfluidics. By parallelizing flow focusing generation units coupled with flow resistors in solvent-resistant materials, microfluidic generation of droplets, bubbles, microcapsules and more recently lipid nanoparticles with precisely controlled size and morphology has been achieved at the liters-per-hr scale. Critical issues in enabling scale-up and outstanding challenges in future development of the scale-up devices will be discussed.

How Responsible Supply Chain Plays a Role in Sustainable Ingredient Design

CONFERENCE

Conference on Engineering Cosmetics and Consumer Products

YEAR

2024

PROCEEDING

5th Engineering Cosmetics and Consumer Products Conference

GROUP

General Submissions

SESSION

Session 1 - Cross-Industry Innovation in Formulated Products

TIME

Tuesday, April 23, 2024 - 9:00am to 9:25am

Authors

Winterburn, S. - Presenter, Rowan University

Inolexâ€™s three pillars of business are innovating sustainably, operating purposefully, and engaging respectfully. The first pillar relates to green chemistry and life cycle thinking which are integral in designing ingredients with a net-positive impact on our world, and inside of the second pillar is the principle of responsible procurement. For our numerous plant-based ingredients, we consciously choose the feedstocks that adhere to green chemistry principles and align with the ethical and sustainability standards that the personal care industry upholds. In this talk I will review some of our key feedstocks like brassica, palm, and castor, and how we take different approaches to mitigate risk in the supply chain, increase traceability, and participate in actions to improve environmental impacts. There are also strategies that can be implemented like regional manufacturing that contribute to overall sustainability by minimizing the distance that products need to travel to get to the end customer. The upstream development of the supply chain as it pertains to feedstock selection, ingredient manufacture, and ingredient transportation is a vital part of maintaining a sustainable material life cycle.

Shifting Behaviors for Laboratory Sustainability

CONFERENCE

Conference on Engineering Cosmetics and Consumer Products

YEAR

2024

PROCEEDING

5th Engineering Cosmetics and Consumer Products Conference

GROUP

General Submissions

SESSION

Session 1 - Cross-Industry Innovation in Formulated Products

TIME

Tuesday, April 23, 2024 - 9:25am to 9:50am

Authors

Ruiz, C. - Presenter, My Green Lab

Every day, laboratories worldwide consume vast resources and generate significant waste, contributing to the climate crisis. This presentation delves into the hidden environmental footprint of laboratory practices and explores how impactful behavioral changes can pave the way toward a more sustainable future. It highlights researchers and lab managers' crucial role in advocating for and implementing environmental improvements.

This session will explore practical strategies for reducing waste, conserving energy, and minimizing chemical use in labs, showcasing that sustainability and scientific advancement can go hand in hand. By embracing a culture of sustainability, laboratories not only benefit the planet but also enhance their efficiency and innovation. Through case studies and evidence-based practices, participants will gain insights into actionable steps for integrating sustainability into their operations, ultimately contributing to a greener, more sustainable future.

Training the Next Generation of Microbiologists: Collaboration is Key

CONFERENCE

Conference on Engineering Cosmetics and Consumer Products

YEAR

2024

PROCEEDING

5th Engineering Cosmetics and Consumer Products Conference

GROUP

General Submissions

SESSION

Session 1 - Cross-Industry Innovation in Formulated Products

TIME

Tuesday, April 23, 2024 - 9:50am to 10:15am

Authors

Loewenstein, M. - Presenter, Q Laboratories

Companies have historically, at their peril, relegated new microbiologists to siloed functions. Whether that be in the QC laboratory, R&D, QA, or elsewhere, new microbiologists build important, but limited expertise, that does little to make them effective in delivering products of high microbial integrity to the marketplace. Only after many years in the industry, perhaps working in several different functions, does the average personal care microbiologists reach a point in their career journey where they can start to see the full picture of holistic microbial control. Microbiologists of the future will face new and varying challenges including how to support sustainability efforts while maintaining high microbial integrity. By proactively outlining how we can combine the expertise of industry specialists, with the reach and institutional resources of academia, the training of microbiologists can be more efficient. With focused collaboration, the next generation of microbiologists can, even as newcomers to the industry, be prepared with the skills and knowledge necessary to be effective on day one.

Effects of Additives on the Flow Behavior of Lamellar-Structured Concentrated Surfactant Solutions

CONFERENCE

Conference on Engineering Cosmetics and Consumer Products

YEAR

2024

PROCEEDING

5th Engineering Cosmetics and Consumer Products Conference

GROUP

General Submissions

SESSION

Session 2 - Engineering Solutions for Sustainable Production

TIME

Tuesday, April 23, 2024 - 11:35am to 12:00pm

Authors

Kelkar, P. - Presenter
Kaboolian, M., Purdue University
Lindberg, S., Procter & Gamble
Williams, E., Purdue University
Erk, K. A., Purdue University

Consumer cleaning products including shampoos and detergents often contain a large amount of water, a critical component typically required at all stages of the product's lifecycle. The water we use in the shower is generally sufficient to quickly dilute personal care products and use them effectively. Changing their consistency from a slow-flowing liquid to a thick paste,

effectively concentrating them, would have significant economic and environmental benefits.
Making concentrated formulations, however, is more complicated than simply removing as much water as desired. Conventional production methods cannot process concentrated surfactant-based materials without altering their microstructure and potentially compromising material performance. Unless multiscale material relationships for concentrated feedstocks are developed, industry will continue to waste valuable resources on inefficient and unsustainable manufacturing processes.

In this study, material relationships at low temperatures were determined for concentrated surfactant solutions using a combination of rheological experiments, cross-polarized microscopy, calorimetry, and SAXS. A lamellar structured 70 wt.% solution of sodium laureth sulfate in water was used as a model system. At cold temperatures, the formation of surfactant crystals resulted in extremely high viscosity. The bulk flow behaviour of multi-lamellar vesicles and focal conic defects in the lamellar phase was similar. Shear-induced crystallization at temperatures higher than the equilibrium crystallization temperature range resulted in an unusual complex viscosity peak. The effects of processing-relevant parameters including temperature, cooling time, and applied shear were investigated. Further, industrially relevant additives â€“ sodium chloride, propylene glycol and cetostearyl alcohol â€“ were separately added in varying concentrations (0.25-20 wt.%) and structure-property relationships were experimentally established. Sodium chloride and propylene glycol behaved as desiccants and plasticizers, respectively, while the fatty alcohols increased the surfactant bilayer thickness. Outcomes from this study help answer unexplained processing behavior for concentrated feedstocks and will have implications in improving the sustainability of consumer products.

Holographic Characterization of Biodegradable Poly Lactic-Co-Glycolic Acid (PLGA) Microbeads for Applications in Cosmetics

CONFERENCE

Conference on Engineering Cosmetics and Consumer Products

YEAR

2024

PROCEEDING

5th Engineering Cosmetics and Consumer Products Conference

GROUP

General Submissions

SESSION

Session 2 - Engineering Solutions for Sustainable Production

TIME

Tuesday, April 23, 2024 - 11:10am to 11:35am

Authors

Philips, L. - Presenter
Cheong, F. C., Spheryx, Inc.
Ruffner, D. B., Spheryx, Inc.
Markus, T., Spheryx, Inc.
Lumer, J., Spheryx, Inc.

Microbeads are essential elements of many cosmetic products for opacifying, mattifying, determining the feel of the product, and enabling specific textures. Historically, microplastics have served these needs. More recently, concerns about the environmental impact of microplastics has led to the exploration of alternative materials for microbeads, with emphasis on biodegradability. In this work, Total Holographic Characterization (THC) is used to study Poly Lactic-*co*-Glycolic acid (PLGA) biodegradable microbeads in the size range of 500 nm - 5 Î¼m. THC uses inline holograms to provide quantitative size, refractive index and concentration of colloidal particles. PLGA polymers are FDA approved biocompatible and biodegradable material for biomedical usage and a possible alternative to current polymer-based microbeads for cosmetic applications. THC provides quantitative measurement of PLGA colloidal particle size, refractive index and concentration under various environmental conditions. Our studies demonstrate the effectiveness of THC in differentiating PLGA microbeads from other types of microbeads. Different polymers can be distinguished by their refractive index. We analyze mixtures of PLGA in the presence of microbeads of other polymeric materials, to determine the concentration of multiple components in a heterogeneous mixture in a single measurement.

Predicting the thermodynamics of formulations via machine learning-enhanced theoretical tools

CONFERENCE

Conference on Engineering Cosmetics and Consumer Products

YEAR

2024

PROCEEDING

5th Engineering Cosmetics and Consumer Products Conference

GROUP

General Submissions

SESSION

Session 2 - Engineering Solutions for Sustainable Production

TIME

Tuesday, April 23, 2024 - 10:45am to 11:10am

Authors

Gartner, T. - Presenter

Characterizing the structure and stability of a formulation is a key challenge in the consumer products industry. Especially when considering a shift to more â€˜sustainableâ€™ products, the ability to easily predict how the incorporation of new component(s) affects soft materialsâ€™ structure and properties would be a key enabler of sustainable product development. Polymer reference interaction site model (PRISM) theory is a powerful computational technique to efficiently predict the structure and thermodynamics of liquid-like polymer and soft matter systems. However, the use of PRISM theory can pose practical challenges due to issues with accuracy and numerical stability for some classes of systems. In this talk, I describe our efforts to incorporate machine learning (ML)-based approaches to improve the applicability of PRISM theory to realistic multicomponent systems. Our preliminary results show that the ML-enhanced PRISM theory performs favorably in comparison to widely-used theoretical approaches. We then applied the ML closure to model the results of small-angle neutron scattering experiments on polymer solutions, again showing superior performance compared to traditional approaches. I will close with a discussion of how such combined theoretical-experimental tools could be applied to the development of the next generation of sustainable consumer products.

Analytical Techniques for Sustainability Compliance

CONFERENCE

Conference on Engineering Cosmetics and Consumer Products

YEAR

2024

PROCEEDING

5th Engineering Cosmetics and Consumer Products Conference

GROUP

General Submissions

SESSION

Session 3 - Regulations, Safety, and Compliance

TIME

Tuesday, April 23, 2024 - 1:40pm to 2:05pm

Authors

Li, X. - Presenter, National University of Singapore

Plastic products play a crucial role in various aspects of our daily lives, offering undeniable societal benefits. However, the widespread use and continuous buildup of microplastics (MPs) have raised significant concerns for global ecosystems and human health. Both synthetic and biopolymers contribute to plastic pollution.
Efforts have been intensified at both municipal and research levels to address the issue of microplastics in ecosystems. Municipal initiatives focus on waste management, wastewater treatment, and regulations targeting single-use and industrial applications of microplastics. Simultaneously, research endeavors are dedicated to technological advancements, such as the development of biodegradable polymers and bioremediation techniques to combat generated

microplastics. Biodegradable plastics, being more susceptible to microbial activity, degrade faster than conventional synthetic plastics, offering an ecological alternative. Bioremediation involves using microorganisms to eliminate microplastics, presenting a sustainable approach.

The combination of biodegradable plastics and bioremediation emerges as a promising strategy to mitigate the challenges associated with microplastic pollution in ecosystems. To comprehensively assess the sustainability of microplastics, various analytical techniques, including visual screening, scanning electron microscopy (SEM), Raman spectroscopy, FTIR, etc., are essential. This presentation will demonstrate how these analytical tools can be effectively utilized to ensure sustainability compliance in managing microplastic-related concerns.

Cleaning and Microbial Control within Cosmetic, Consumer Products, and Topical Drug Manufacturing Facilities

CONFERENCE

Conference on Engineering Cosmetics and Consumer Products

YEAR

2024

PROCEEDING

5th Engineering Cosmetics and Consumer Products Conference

GROUP

General Submissions

SESSION

Session 3 - Regulations, Safety, and Compliance

TIME

Tuesday, April 23, 2024 - 2:05pm to 2:30pm

Authors

Lopolito, P. - Presenter, STERIS

Global cGMP regulations for cosmetic and drug products require control of the cleaning and sanitizing practices within the manufacturing process to ensure contaminates do not adversely impact the quality of the products produced. Contaminates can include active pharmaceutical ingredients, non-active ingredients, dyes, fragrances, cleaning and sanitizing agents, microbial organisms, endotoxins, exotoxins, and prions. The acceptable level of contaminates will depend on the contaminate, nature of the product, storage conditions, route of administration and the intended recipient of the product. Laboratory testing is the best way to design successful cleaning procedures.

Selection of the sanitizing agent and controlling critical parameters is the best way to ensure success. The cleaning validation *life cycle* concept changes the emphasis from validation to design and monitoring of the cleaning and sanitizing practices. Design and monitoring of critical parameters ensure control of processes. Monitoring of the process also encourages improvements through change management and real-time scientific-based decisions to out of specification (OOS) results. The presentation will also share cleaning and microbial challenges, such as *Burkholderia,* as manufacturing facilities improve cleaning and sanitizing practices.

MoCRA Explained: Understanding New FDA Cosmetic Regulations

CONFERENCE

Conference on Engineering Cosmetics and Consumer Products

YEAR

2024

PROCEEDING

5th Engineering Cosmetics and Consumer Products Conference

GROUP

General Submissions

SESSION

Session 3 - Regulations, Safety, and Compliance

TIME

Tuesday, April 23, 2024 - 2:30pm to 2:55pm

Authors

Bellomo, J. - Presenter, Registrar Corp

In December 2022, U.S. Congress passed a bill that will forever change the FDAâ€™s authority over cosmetic regulations in the United States. The Modernization of Cosmetics Regulation Act of 2022 (known as â€œMoCRAâ€) focuses on provisions that will require domestic and foreign firms to register with US FDA, designate a US Agent for FDA communications, â€œlistâ€ their products, collect and report serious adverse events to US FDA, and many other transparency and safety elements.
With 2024 set to provide more insight into other regulations around Good Manufacturing Practices, Fragrance Allergens, and Labeling requirements, it is important to understand the compliance deadlines and requirements and how they apply to your business.

Though there are still many unknowns, there are steps companies can take now to prepare and educate themselves on what these changes mean to their business. In this seminar, it will outline what the new requirements mean to facilities and Responsible Persons, and how this will transform the industry for the years ahead.

Why Microbiological Contaminations are the Largest Cause of Cosmetic Recalls, and How to Avoid One.

CONFERENCE

Conference on Engineering Cosmetics and Consumer Products

YEAR

2024

PROCEEDING

5th Engineering Cosmetics and Consumer Products Conference

GROUP

General Submissions

SESSION

Session 3 - Regulations, Safety, and Compliance

TIME

Tuesday, April 23, 2024 - 1:15pm to 1:40pm

Authors

Waby, G. - Presenter, Obelis USA

According to FDA figures, the single largest reason for recalls of Cosmetics, Drugs and Dietary Supplements is microbiological contamination. The development of new more â€œnaturalâ€ formulations, loss of robust preservatives and legacy manufacturing and processing are key drivers of this trend. Hygienic manufacturing requires the coordination of formula development, preservation, process design, scale-up, facility and equipment design, operations, training, quality assurance and testing, as well as leadership at all levels, to minimize risk.

Hear how this has been addressed by a task force from industry, sharing knowledge and best practices, the role of associations, similar to AIChE, academia and individual companies. Congress passed the Modernization of Cosmetics Regulations Act at the end of 2022. According to this regulation, the FDA must publish draft guidance for industry on Good Manufacturing Practices during 2024 and finalize them in 2025. The Act greatly increases the level of scrutiny of the cosmetics industry, including the ability to control microbiological risk.

A Naturally Great Hair Day: A sustainable alternative to quaternary compounds

CONFERENCE

Conference on Engineering Cosmetics and Consumer Products

YEAR

2024

PROCEEDING

5th Engineering Cosmetics and Consumer Products Conference

GROUP

General Submissions

SESSION

Session 4 - Novel Materials in Hair Care

TIME

Tuesday, April 23, 2024 - 3:50pm to 4:15pm

Authors

de Szalay, S. - Presenter, Gattefossé

As the consumer need for natural, clean, and sustainable hair care products continues to grow, greener alternatives are a key need for quats and silicones. Natural oils and waxes are known but can create a greasy appearance and feel with minimal volumizing benefits. A natural, patented alternative to silicones and quats based on functionalized jojoba and sunflower seed waxes is presented. The technical origin, background and molecular composition of this material is discussed and demonstrates its biomimetic nature and similarity to human sebum.

This new material was benchmarked against traditional quats and silicone oils. Technical evaluations include contact angle, volume retention, curl retention, contact angle, frizz, wet and dry comb, and sensorial properties. Data showing a statistically significant increase of hair volume reduction in humid conditions is presented. Hair frizz was also significantly reduced and curl retention was also maximized after exposure of hair tresses to humid conditions versus quats and silicones. In wet hair combing, statistically significant wins over silicone oils were achieved. In dry combing and hair breakage, the performance of this material rivaled that of both quats and silicones. Anti-stat benefits were also achieved with synergies observed when combining with quats. Finally, formulations tested with this new material were positively rated by consumers.

In summary, this new ingredient with a bio-similar quality to sebum not only rivals the performance of quats and silicones but exhibits a more desirable biodegradability profile.

Designing Functional Polysaccharides for Formulated Personal Care Products

CONFERENCE

Conference on Engineering Cosmetics and Consumer Products

YEAR

2024

PROCEEDING

5th Engineering Cosmetics and Consumer Products Conference

GROUP

General Submissions

SESSION

Session 4 - Novel Materials in Hair Care

TIME

Tuesday, April 23, 2024 - 3:25pm to 3:50pm

Authors

Leal, L. - Presenter, The Dow Chemical Company

The home and personal care market has been experiencing revolutionary demand for sustainable alternatives to traditional synthetic materials used across market segments, particularly in rinse-off products that consumers see going down the drain and materials placed directly on their skin. While consumer pressure is gaining momentum for sustainable materials, there is no tolerance for a decrease in performance. When consumers use a shampoo, they are expecting a certain level of cleaning performance, but also expect a level of conditioning that provides them with brushable hair, both wet and dry. Similarly, with a rinse-off conditioner, hair should feel silky and smooth, and any perception of damage should be reduced. Dow Home and Personal Care (HPC) explored sustainable alternatives that provide conditioning from both shampoo and rinse-off conditioner formats. This talk will share our journey to sustainable materials for shampoo and rinse-off conditioner products and showcase our solutions for the deposition of actives and conditioning properties on hair.

Maizecare Clarity Polymer (Hydrolyzed Corn Starch): Sustainable Alternative to Synthetic Styling Polymers.

CONFERENCE

Conference on Engineering Cosmetics and Consumer Products

YEAR

2024

PROCEEDING

5th Engineering Cosmetics and Consumer Products Conference

GROUP

General Submissions

SESSION

Session 4 - Novel Materials in Hair Care

TIME

Tuesday, April 23, 2024 - 4:15pm to 4:40pm

Authors

Joshi, K. - Presenter

Hair care consumers today are increasingly focused on selecting products which contain natural ingredients without sacrificing performance. The current incumbent product is a synthetic, petroleum-based polymer styling aid of polyvinylpyrrolidone (PVP), with a few niches natural products for customers valuing sustainability over performance. PVP provides excellent hair gel transparency and hair stiffness but lacks curl retention performance. Hair styled with PVP collapses when exposed to high humidity. Of greater concern, PVP is under scrutiny as a suspected carcinogen and reproductive toxicant under ECHA Annex III, which is of particular concern for a personal care product and one that â€œgoes down the drainâ€ . MaizeCareâ„¢ Clarity Polymer is the first product to provide the sustainability profile of natural products and exceed the performance of synthetics, setting a new standard. It comes from a sustainable feedstock and has a CO_2 footprint 78% lower than that of PVP.

Fundamental properties such as water solubility/dispersibility, film formation, and modulus determine the performance of a hair fixatives. Structure-property-performance studies were complete to identify the best natural material candidates. Optimum molecular weight of MaizeCareâ„¢ Clarity Polymer was identified to provide the optimal hair stiffness to gel clarity balance. If the molecular weight is too low, the hair gel will not provide good stiffness. If the molecular weight is too high, the hair gel will not be clear. In addition, the polymer does not leave a white residue or flaking on the hair after combing. MaizeCareâ„¢ Clarity Polymer provided greater than 90% curl retention due to the non-hygroscopic nature of the polymer, which is far superior compared to polyvinylpyrrolidone (PVP). MaizeCareâ„¢ Clarity Polymer can be easily formulated for all hair types into various product formats & providing the creative textures consumers demand with a customized consumer experience.

Carboxymethyl Dextran As Biodegradable Dispersants for Automatic Dishwashing Detergents

CONFERENCE

Conference on Engineering Cosmetics and Consumer Products

YEAR

2024

PROCEEDING

5th Engineering Cosmetics and Consumer Products Conference

GROUP

General Submissions

SESSION

Session 5 - Sustainability in Home Care

TIME

Wednesday, April 24, 2024 - 10:10am to 10:35am

Authors

Yin, L. - Presenter
Mercando, P., Dow Chemical
Pulukkody, R., Dow Chemical
Backer, S., Dow Chemical
Cauvin, S., Dow Chemical
Mitchell, M., Dow Chemical
Scoles, M., Dow Chemical
Albright, V. III, Dow Chemical
Jing, M., Dow Chemical
Reiner, B., Dow Chemical
Upshur, M. A., Dow Chemical

Volz, C., Dow Chemical

Polymeric dispersants are additives commonly used in household cleaning products such as automatic dishwashing (ADW) and laundry detergents. Incumbent technologies such as polycarboxylates and their copolymers offer excellent cleaning performance but lack biodegradability. Therefore, these materials after use often linger in the natural environment for years to decades. Biodegradable alternatives are highly desired for these â€œdown-the-drainâ€ applications which would reduce environmental persistence and enable carbon circularity. Poly(aspartic acid) and carboxymethyl inulin show biodegradability but so far have limited market penetration given their unsatisfactory cleaning performance. Here we report the synthesis of carboxymethyl dextran and their use as biodegradable, high-performing dispersants for ADW detergents. Dextran is a linear, water-soluble poly(Î±-glucan) derived from the fermentation of sucrose from natural sources such as sugar cane. Carboxymethyl dextran was synthesized by the caustic-activated alkylation with sodium chloroacetate in water. ADW detergents with carboxymethyl dextran showed excellent anti-spotting and anti-filming performance in model citrate-based formulations. The structureâ€'performance relationship for optimized compositional design (e.g., degree of substitution and molecular weight) will be discussed.

Creating a sustainable ecosystem for producing surfactant and other ingredients

CONFERENCE

Conference on Engineering Cosmetics and Consumer Products

YEAR

2024

PROCEEDING

5th Engineering Cosmetics and Consumer Products Conference

GROUP

General Submissions

SESSION

Session 5 - Sustainability in Home Care

Authors

Ontumi, D. - Presenter, Ruby Bio

Every day, laboratories worldwide consume vast resources and generate significant waste, contributing to the climate crisis. This presentation delves into the hidden environmental footprint of laboratory practices and explores how impactful behavioral changes can pave the way toward a more sustainable future. It highlights researchers and lab managers' crucial role in advocating for and implementing environmental improvements.
This session will explore practical strategies for reducing waste, conserving energy, and minimizing chemical use in labs, showcasing that sustainability and scientific advancement can go hand in hand. By embracing a culture of sustainability, laboratories not only benefit the planet but also enhance their efficiency and innovation. Through case studies and evidence-based practices, participants will gain insights into actionable steps for integrating sustainability into their operations, ultimately contributing to a greener, more sustainable future.

Energy saving solutions for home care products

CONFERENCE

Conference on Engineering Cosmetics and Consumer Products

YEAR

2024

PROCEEDING

5th Engineering Cosmetics and Consumer Products Conference

GROUP

General Submissions

SESSION

Session 5 - Sustainability in Home Care

TIME

Wednesday, April 24, 2024 - 9:45am to 10:10am

Authors

Kamin, S. - Presenter, Ashland

To tackle climate change requires effort from every one of us, at the individual level. Therefore, everyday use products should be designed to allow consumers to change their habits towards more sustainable ones. Solid formats provide several sustainable benefits, reducing the carbon footprint of transportation, allowing for paper-based packaging and reducing waste with single dose units. The challenge of concentrating nature derived and biodegradable ingredients while delivering outstanding performance and sensory is exemplified with the powder to liquid hand dish formulation containing Cellulose Gum and a guar derivative. Improving the performance of a product for optimal cleaning at lower dosage is shown with a glass cleaner tablet that transforms to a liquid. In addition, the energy costs being on the rise, novel technologies should focus on energy saving solutions, thus providing both benefit for the planet as well a tangible saving for the consumer. Powder laundry detergents that give good cleaning performance at low temperature can be designed thanks to a biodegradable soil release agent.

Alpha-Tocopherol Microcapsules Added with Moringa Oleifera Supercritical CO2 Extract As a Release System for Transdermal Applications

CONFERENCE

Conference on Engineering Cosmetics and Consumer Products

YEAR

2024

PROCEEDING

5th Engineering Cosmetics and Consumer Products Conference

GROUP

General Submissions

SESSION

Session 6 - Novel Materials in Skin Care

TIME

Wednesday, April 24, 2024 - 11:00am to 11:25am

Authors

Kessler, J. - Presenter
Manrique, Y. A., Faculdade de Engenharia da Universidade do Porto - FEUP
Martins, I., LSRE-LCM-Laboratory of Separation and Reaction Engineering-Laboratory of Catalysis and Materials, Faculdade de Engenharia, Universidade do Porto
Rodrigues, A. E., LSRE - Laboratory of Separation and Reaction Engineering - Associate Laboratory LSRE/LCM
Barreiro, F., Centro de Investigação de Montanha (CIMO), Instituto Politécnico de Bragança
Dias, M. M., LSRE - Laboratory of Separation and Reaction Engineering - Faculty of Engineering - University of Porto

Introduction:

The design of the coacervate systems is proposed to ensure the bioavailability enhancement of cosmetic ingredients, attending to the demand for nourishing, anti-ageing, and, more recently, medical drug-like formulations [1]. *Moringa oleifera* L. leaf extract (Mo), obtained by supercritical fluid extraction with dioxide carbon (SFE-CO_2), was found to be a suitable candidate for cosmetic uses due to its rich Î±-tocopherol composition [2].

Built upon developing modelling microencapsulation by complex coacervation (MCC) using Arabic gum and gelatin adaptable to enclose numerous cosmetic ingredients, this study introduces the microencapsulation of Mo combined with Î±-tocopherol (Mo+Î±-toc) and Î±-tocopherol alone (Î±-toc). The aim is to explore their diffusion into synthetic membranes resembling human skin.

Methodology:

Physicochemical and imaging analyses were applied to the resulting products. Particle size and shape confirmed the morphology of the microcapsules, while the encapsulation efficiency (EE%), loading content (LC%), and solid content (SC%) measured the process viability.

The microcapsules and non-coacervated bioactives were also submitted to diffusion tests in a Franz cell device using synthetic Strat-M transdermal membranes to emulate human skin.

Results:

Imaging analyses depicted spherical delivery systems with a multinucleated morphology featuring bioactives within a porous structure covered by a dense polymeric shell. The average particle size was circa 60 Î¼m. High encapsulation efficiency was confirmed, reaching 93.4Â±1.3% and 95.1Â±1.2% for Mo+Î±-toc and Î±-toc, respectively, with no significant differences. LC% was 10.0Â±1.3% and 10.3Â±1.2%, respectively, in which the load could reach no more than 14%, and SC% were determined within 8.3Â±0.1% and 8.4Â±0.5%.

Franz cell studies revealed faster diffusion of bioactive when incorporated in the microcapsules compared to non-coacervated bioactive, with receptor fluid and membrane showing significantly higher Î±-tocopherol content for microencapsulated forms.

References:

[1]Kessler, J. C., et al. (2023). Separations, 10(3), 210. https://www.mdpi.com/2297-8739/10/3/210

[2]Taofiq, O., et al. (2018). Industrial Crops and Products, 124, 44-52. https://doi.org/https://doi.org/10.1016/j.indcrop.2018.07.057

Biobased Glycolic Acid Is a Suitable One-to-One Replacement for Synthetic Glycolic Acid in Skincare Formulated Products

CONFERENCE

Conference on Engineering Cosmetics and Consumer Products

YEAR

2024

PROCEEDING

5th Engineering Cosmetics and Consumer Products Conference

GROUP

General Submissions

SESSION

Session 6 - Novel Materials in Skin Care

TIME

Wednesday, April 24, 2024 - 11:50am to 12:15pm

Authors

Curry, A. - Presenter, Good Molecules, LLC
Sahloff, K. - Presenter, Good Molecules, LLC
Gallagher, E. - Presenter, Good Molecules, LLC
da Silva Souza, I.

Glycolic acid (GA) has a seasoned history as an effective skin exfoliating agent. Skincare alone accounted for 40% of the $468.2 million global GA market size in 2019[1]. As personal care consumers and manufacturers look for more sustainable formulations, research into ingredient sources and manufacturing processes has become necessary. Herein, we report a direct comparative analysis of a synthetic 70% GA aqueous solution with a biobased, fermentation-derived alternative. Titration and chromatographic methods were used to determine water, GA, and impurity content or chemical identity. A 3-month accelerated shelf-life study (RT, 37Â°C/75% R.H., and 50Â°C/35% R.H.) with versions of the Good Moleculesâ€™ Overnight Exfoliating Treatment was conducted to compare the effects of formulating with either the synthetic or biobased material. Negligible differences were found in each solution's total acidity and water content, as measured by potentiometric and Karl Fisher titrations, respectively. Sample purity determined by ion exclusion chromatography (IEC) showed a major (â^¼10%) impurity peak in both samples and a minor additional impurity peak (â^¼0.1%) in the biobased GA. Further, GC-MS analysis showed that Diglycolic Acid was the major contaminant in both samples. Interestingly, Glycine was resolved only from the biobased GA solution. As with most amino acids, the residual Glycine found in the biobased GA could potentially offer additional skincare benefits. The accelerated stability study showed no meaningful changes in the formulated products' appearance, odor, color, viscosity, or pH. Overall, biobased GA offers a major opportunity for sustainable improvement in product development as an easy-to-formulate, one-to-one synthetic GA replacement.

[1]Fortune Business Insights (2020, October). *Glycolic Acid Market Size, Share & COVID-19 Impact Analysis [...] and Geography Forecast, 2020-2027*. https://www.fortunebusinessinsights.com/industry-reports/glycolic-acid-market-101922

Sustainable Products for Skin Care and Cosmetics

CONFERENCE

Conference on Engineering Cosmetics and Consumer Products

YEAR

2024

PROCEEDING

5th Engineering Cosmetics and Consumer Products Conference

GROUP

General Submissions

SESSION

Session 6 - Novel Materials in Skin Care

TIME

Wednesday, April 24, 2024 - 11:25am to 11:50am

Authors

Xu, W. - Presenter

Two sustainable products with optical effects were developed for skin care and cosmetic markets, aiming to address the evolving consumer and regulatory needs for eco-friendly and performing materials. These innovations include an ECOCERT-approved silica naturally derived from upcycled rice husk and a bio-based and readily biodegradable SPF booster with 100% renewable carbon.

The rice husk is made through a particle technology engineered to enable excellent optical soft-focus effects to reduce the appearance of skin blemishes and sebum absorption. The particle size was optimized to enable efficient visible light scattering and desired sensorial benefits.

The SPF booster, SunSpheresâ„¢ BIO, is a specific crystalline cellulose-based material made with a patented technology of wood hydrolysis. It`s designed to avoid performance trade-offs commonly exhibiting in â€œgreenâ€ chemicals and address the key need in Suncare industry. Sun/ultraviolet (UV) exposure is the primary cause of non-biological skin cancer and skin aging. The particle size and the refractive index (1.51-1.56) of this material lead to efficient UV scattering. Such scattering deviates sunlight from its original path and increases the optical path length of the UV inside the sunscreen film, resulting in a higher probability for the UV filters to absorb or back-scatter the harmful UV radiation. This eco-friendly product therefore maximizes the skin protection from sun/UV damage while minimizing the negative health and environmental impact of sun care products. It provides additional rheological benefits such as water thickening and excellent shear thinning for spreadability and sprayability. The ingredient can be easily formulated into a variety of formulation chassis, including oil in water, water in oil, and water in silicone, with either organic or inorganic UV filters. This product received several global awards, including *In-Cosmetic® Global- Silver Award of Innovation Zone Best Ingredient 2020* and *ICIS Innovation Award 2021- Product with Best Benefit to the Environment and Sustainability.*

AUTHOR INDEX

AIChE
120 Wall Street, FL 23
New York, NY 10005-4020

ISBN 979-8-3313-0080-7

9 798331 300807

CURRAN ASSOCIATES INC.
proceedings.com